A Cultural Paradox:
Fun in Mathematics

12/13/09

Dear Jamie,

Enjoy the Maths!

Your Cousin,

Jeff

Jeffrey A. Zilahy

A Cultural Paradox:
Fun in Mathematics

J 2 the Z Publishing

Copyright © 2009 by Jeffrey A. Zilahy

Published by Lulu.com

Imprint: J 2 the Z Publishing

All rights reserved. Printed in the United States of America. Perfect Bound Paperback

ISBN 978-0-557-12264-6

Dedicated to 40

Acknowledgements

I thank everyone that has ever inspired me. My family, my parents, my sister and brother-in-law, those people I am lucky to call true friends, my math guru compadres Anderson and Kolenda, my grandpa Zoltan and my grandma Gabriella for being totally rad and Zen-like, my Business Calculus class Professor Zeitz for making me see mathematicians as cool, Dr. Lombardi, my Calculus 1 teacher, for being one of my best teachers even if I have never met him in person, my Vector Calculus tutor Sid, my college Writing Professor Kevin DiPirro who implored me to never stop writing, my friend and mentor Leon Thomas, and all those bright young minds that I have had the chance to call my students. I also must give thanks to anyone that challenges traditional ways of thinking *, people like Stephen Wolfram, Ray Kurzweil, Michio Kaku, Eliezer Yudkowsky, and Cliff Pickover. You have all helped to keep me at the forefront of wonder and awe!

Jeff

October 12th 2009

I must tip my hat to those monkeys using nothing more than a typewriter and infinite time and were able to type up this exact book, and without a single typo!

Book Directory:

1. Introduction

2. Picking a winner is as easy as 1, 2, 3.

3. That's my Birthday too!

4. Sizing up Infinity

5. I am a Liar

6. Gratuitous Mathematical Hype

7. LOL Math, Math LOL

8. Why High School Geometry is not the full story

9. Abstraction is for the Birds

10. NKOS: Anti-Establishment might be the Establishment

11. 42% of statistics are made up!

12. Undercover Mathematicians

13. $e^{i*Pi} + 1 = 0$ is heavy duty

14. Economic implications of Gaussian Copula Functions

15. A Proven Savant

16. E8 and the history of the TOE

17. One heck of a ratio

18. A real Mathematical Hero

19. Casinos love Math

20. The Man who was sure about Uncertainty

21. More Incompleteness

22. Have I got a Question for you!

23. When Nothing is Something

24. Think Binary

25. Your Order will take forever

26. When you need Randomness in your life

27. $e=mc^2$ Redux

28. From the Quipu to Mathematica

29. Through the eyes of Escher

30. Origami is realized geometry

31. Quantifying the Physical: Metrics in Sports

32. Geometric Progression sure adds up

33. Nature = a + bi and other Infinite Details

34. Mundane implications of Time Dilation

35. Off on a Tangent

36. Modern Syntax Paradigms

37. Awesome Numb3rs

38. A fun sampling of Math Symbols

39. Your consciousness can be computed

40. Auto-didactic Ivy Leaguers

41. Zeno's Paradox in time and space

42. I needn't say anymore

43. I can see the past as it were

44. Music is a beautiful triangulation

45. Mobius Strip: assembly required

46. I will never use this

47. Choice words

48. Latest in building marvels

49. Your eyes do not tell the whole story

50. This kinda Internet Cred is worth paying attention to

51. The A.I. Inflection Point

52. I know Kung Fu

53. We eat this stuff up

54. Alpha behavior

55. That thing on your wrist is just a Temporal Dimension Gauge.

56. Stream of consciousness, Bing for details

57. Q.E.D.

Bibliography Note & Index

CH 1: Introduction

To begin, I readily confess to being an unabashed science and technology aficionado. It is this passion and excitement that propelled me in writing this book; I did so from August into October 2009. I have always loved the objective nature of science and been amazed at how much it has wrought for the inhabitants of this little rock we call Earth. I seem to repeatedly find myself reminded of the pure chance of being born in this era and how awesome it is to be involved in science in such a time. I am of the belief that mathematics is the underlying "software" that powers the "hardware" that we live in, namely planet earth and of course the greater cosmos. It seems there is a direct relationship between the level of sophistication that a civilization is at and the degree of mathematics that a civilization grasps. I would say that math is powerful, intriguing, and intensely relevant to all of our lives! The purpose of this book is definitely not meant to be an exhaustive look at mathematics. It is not meant to be in depth on any subject that is covered or to be written using a bunch of technical jargon. Rather, it is written as a reflection on what I consider some of the most interesting and powerful ideas that swirl around in the mathematics community today. Really, this book represents my musings on early 21st century recreational maths. Naturally, there are countless other topics and areas of mathematics that can be addressed. The topics in this book are the ones that are on my radar right now and that are tied into our popular culture. I also must confess that while this book is classified as Non-Fiction, and therefore the ideas herein are presumed to be fact, I have tried my best to be accurate, so please pardon any of my mistakes.

"The last thing one knows when writing a book is what to put first."

Blaise Pascal

As it pertains to the sequencing of the book, the chapters do not follow any set order, and part of that purpose is to allow the reader to jump in anywhere in the book, at any time, and get a new topic in a couple pages of reading. The only real thread that ties all the chapters together is they are all mathematical in nature. Hopefully they are also as interesting and surprising to you as they are to me.

CH 2: Picking a winner is as easy as 1, 2, 3.

The Monty Hall problem is an example of how mathematics can sometimes be counter-intuitive to common sense. It is so named for the game show host, Monty Hall who actually featured this problem on a real live game show. This problem deals with probabilities. The typical set up involves 3 doors. The contestant (i.e. you!) is told that behind 2 of the doors are two undesirable prizes, let's say a working desktop computer running windows Me and with minimum RAM. Behind the 3rd door is a really desirable prize, say the Nissan GT-R. Monty starts by asking which door you believe the Nissan is behind. You say Door 1 since there can be only one prize. He then surprises you by opening door number 2 revealing a giant clunky outdated computer! The audience lets out a gasp as Monty turns to you and asks whether you would like to switch to door number 3. Now the question to you the reader is to ascertain whether you would increase your odds of winning that prized car by switching from door number 1 to door number 3. Most people incorrectly assume that both door 1 and 3 have the same probabilities of revealing the car. In actual fact, the switching from door 1 to 3 is a very wise move. You go from 1/3 chance of finding the car in door number 1 to 2/3 chance of finding the car in door number 3. Why? Well when you first were asked to pick a door, all 3 doors had the same chance of revealing the car. That means whichever door you chose, 1, 2 or 3, 1 in this case, you have a guaranteed in 1/3 chance of getting the right door. Now when Monty opened the surprise door, door number 2 and eliminated that door as an option for containing the car, which means that now you are contending with only 2 doors where you will find the car. But as we said before, your door, number 1, is a 1/3 probability of being the correct door. Since we know that there is a 100% chance of it being either door number 1 or door number 3, and since we also know that door number 1 represents 1/3 of that probability, then we know that the remaining 2/3 must belong to door number 3. So essentially, by revealing door number 2, we increased the probability of the door you did not choose, door number 3! I told you it would be counter-intuitive!

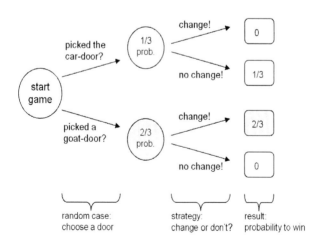

CH 3: That's my Birthday too!

The birthday is not a paradox by definition but more of an unexpected resulting probability of a given event. The idea here is that you are given 23 totally random people. (If you wish to digress into randomness, see chapter 26.) You are told that there are 365 different days on which you can call your birthday; the leap year would count as .25 of a day, occurring once every 4 years. You are then asked what percentage chance (0-100%) there is that any two of these twenty three people maintain the same birthday. The truth is that there is a slightly better than 50% of there being two people with the same birthday! This means chances a bit better then a flip of a coin! Additionally, if you were to have a paltry 57 people to test the experiment, you have the surprising guarantee of more than 99% certainty of matching two birthdays! The reason why this fact is so is because every person's birthday is being compared with everyone else's birthday. This means that if you were to create unique pairs of people from the 23 people, you can obtain 253 different pairs of people. By considering the group of 23 people in terms of the numbers of pairs (253) then the number of individuals (23), the result is less surprising. The mathematics involved does assume a perfect distribution of birthdays, meaning any given birthday has an equal probability of happening with any other given birthday, and in reality certain birthdays are in fact more common than other, this skewing of the data does not affect predicted results when actually trying out the experiment. The math involved in the birthday paradox is also related to the pigeonhole principle. In the pigeonhole principle, if you have n pigeons and p holes and $1 < p < n$, then you are assured that at least one of the holes will contain 2 pigeons. This is the same logic that ensures if you know someone has 3 children, you know that at least 2 of the children are of the same gender. Computer scientists might have also recognized this problem from hashing algorithms, where the number of keys exceeds the number of indices in an array.

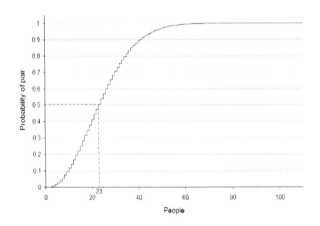

CH 4: Sizing up Infinity

Most people are well acquainted with the word infinity and know it to mean a never ending value, whether it be the debut album from Marshall Mathers, Infinite, or anytime someone wants to evoke a boundless value or idea. In order to understand the idea of infinity strictly in terms of numbers it helps to first look at the idea of a set. In math a set is a collection of objects, of which numbers are often the objects. We also can have no objects, in which case we call it the empty set. Get at the computer scientist rapper MC Plus+ for more information. We could have a set of {1,5,9} or {2,4,6,8,10.....} where the dots indicated that the numbers are going on forever. So here we find ourselves with our first infinite set, the set of even numbers, or to be more rigorous, any number that divided by 2 yields an integer value. Speaking of, now let's consider another infinite set, the set of Natural numbers. This set is 1,2,3,4,5... and on we go for infinity. Now that you have an idea of those numbers marching down a never ending line, consider the set of the Real Numbers. This set includes all the natural numbers, but also irrational numbers like Pi and e (Irrational numbers are numbers that cannot be expressed as the ratio of two integers). Now what is interesting is how it has been proven that the size of numbers between 0 and 1 of the Real numbers is even greater than the entire set of Naturals. Part of the way in which infinities are measured has to do with the idea of correspondence or having a partner element in one set with a partner element in another set. Imagine you have two bags filled with marbles but are not sure which bag has more marbles. All that you need to do is to take a count as you take out one marble from each bag and if all the marbles in each bag are emptied simultaneously then we know there are the same number of marbles in each bag. In this similar way of establishing correspondence, we do with the natural numbers. The result is that there ends up being more Reals than can find matches with the naturals and thus, the infinity of the Reals is far greater than the infinity of naturals. Get at Cantors Transfinite Numbers for more information.

CH 5: I am a liar

Since all the rules behind a language, called axioms in math, are inherently governed with logic, then logical fallacies that appear in our language can be analyzed with mathematics. Let us take a careful look at the following sentence: "This statement is false." What is paradoxical about this is if indeed you accept the statements premise then you are caught in a logical loop where by accepting its premise you are simultaneously rejecting it since the statement is claiming to be false. So false = true and true = false. This makes the statement simultaneously true and false at the same time, which is impossible. Another example of this logical paradox is called Jourdain's Card Paradox. Imagine a card on which one side is written: "The sentence on the other side of this card is true." On the other side is written: "The sentence on the other side of this card is false." As you can tell, you end up in a paradoxical logical loop.

A final example to consider is a card with the following 3 sentences printed:

1. This sentence contains five words.
2. This sentence contains eight words.
3. Exactly one sentence on this card is true.

CH 6: Gratuitous Mathematical Hype

It can become confusing in this hyper-speed, information-overload world of ours to make heads and tails of science and in particular how everything in science is related to one another. Consider the following visual approach to better clarify science. The tree metaphor: the trunk is math; its largest branch is physics, its largest branch off that is chemistry, then biology, then psychology. Clearly, physics is the science of using math to explain the world around us and therefore physics is really, no matter how many physicists might try to argue with mathematicians otherwise, just applied math. ;) While there will forever exist some degree of dispute over which majors in college are the most challenging and difficult, it is hard to argue that the most pure and pervasive science is undeniably mathematics. You could also replace the tree metaphor with a series of ever smaller concentric circles, with math being the outermost circle. Here also is the faux inequality: Math > Physics > Chemistry > Biology > Psychology.

CH 7: LOL Math, Math LOL

A good mathematical joke is better, and better mathematics, than a dozen mediocre papers.

J. E. Littlewood (A Mathematician's Miscellany)

The spectrum of humor that deals with mathematics actually is good at revealing some of the underlying concepts and truths behind universal math concepts. It is also a chance to poke fun at ourselves and laugh a little. Here are several of my favorite math jokes, of which I also have what is likely to be one of the world's biggest collections. A bibliography of said collection can be found at the URL mathematicshumor.com if you are interested in tracking down more math humor, and really, who wouldn't?

Why was the number 6 afraid of its consecutive integer 7? Because 7 8 9!

5q + 5q = ? You are welcome!

There are three kinds of people in the world; those who can count and those who can't.

Did you know that 5 out of every 4 people have a problem with fractions?

Why does 2L - 2L say Christmas?
Noel!

In order to do well in geometry it helps to know all the angles.

The angles that get all the attention are always the acute ones.

What did I overhear the number zero say to the number eight? Very nice belt!

The number Seven is interesting as it is the only odd number that can easily be made even, how?
Seven - s = even

Why is the presence of 2 doctors a strange occurrence? Because it represents a paradox!

What did the little seedling finally say when he was a full grown tree?
G-E-O-M-E-T-R-Y!

What animals are best at maths? Well Rabbits can multiply, fish are always in schools and butterflies are natural mothematicians.

If a mathematician ever opens a gin distillery, call the product OriGin, [0,0]

You might overhear mathematicians at restaurants ordering pairs or pi!

The mathematician's favorite part of the newspaper has to be the conic section.

If you are against the metric system then you really are just a de-feat-ist!

Don't give your math book a hard time, it has its own problems.

A father said to his son as looked over his math homework, "Are schools teachers STILL looking for the lowest common denominator?"

Why are the numbers 1-12 good security?
Because they are always on the watch!

CH 8: Why High School Geometry is not the full story

For those of you that have made it through a high school geometry class, concepts like Pi (Chapter 17), acute and obtuse, parallel and perpendicular are probably all terms that you can recall. Well, one of the central tenants we were all taught as an absolute truth was surrounding a fact about triangles. It stated that the angle sum of any triangle (A 3-gon) must always equal 180 degrees. I don't know about you but the way it was presented to me was this was always the case no exceptions! The truth is that in many situations triangles can actually have more or less than 180 degrees! This is because there are different types of geometric systems. High school geometry is actually formally called Euclidean geometry and refers to the geometry we see with modern architecture, and most of man-made creations. In nature, geometry doesn't always follow such a script. For most of human history, it was assumed that Euclidean Geometry was absolute, there was nothing else to consider. In the past few centuries, we have come to realize there are other geometries that can accurately describe physical space! Two of the most common Non-Euclidean types are called Hyperbolic and Elliptic Geometries. The way in which these geometries differ is in the modification of a central tenant of Euclidean geometry, the Parallel postulate. The reality is that in other geometries parallel lines can cross one another or get further away from one another! Part of this reason is that Euclidean geometry is modeled on the notion of a flat plane. When you consider the Earth as a sphere, in its Geometry, rather aptly named Elliptic, considering Earths somewhat oblong shape, the sum of the angles of a triangle on such a surface is not exactly 180 degrees!

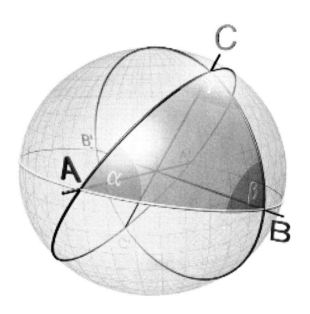

CH 9: Abstraction is for the Birds

The next time you hear someone spout how they are unable to do math, that they lack the "gene", remind them that mathematical thoughts are ingrained in all humans, in fact, in all animals. Still, it took humanity quite a bit of time to get some math problems worked out, the Sumerians invented numbers in 8000 BC and the Greeks made mathematics a centerpiece of their civilization in 600 BC. Today of course, mathematics has been harnessed to an unparalleled level, as is readily evident by modern society. While it is true that not everyone walking around are doing long division or triple integral problems in their heads, we are indeed a very spatially oriented species. This means we are automatically calculating distances, comparing values and doing a great deal of applied math as it relates to our 3 dimensional surroundings. In animals we are continuously discovering new ways they are able to use their own forms of math to survive and adapt to their surroundings. Before we consider a few examples let's begin with some perspective. The likely silver medalist for intelligence on this planet of ours, the chimpanzees, can through intense training attain the skills of….drum roll please….a human two year old. So without any surprise, as far as we know of the animal kingdom to date, no animal is able to abstract and create symbolic language for mathematics anywhere near a human level. Nevertheless, we are constantly learning of how animals are able to adapt to their surroundings through the use of mathematical principles. Consider the crow's logical problem solving ability. In the experiment, a container of water is placed in the crow's environment. The level of the water is intentionally made too low for the crow to access for a drink. The crows have repeatedly demonstrated an ability to use rocks that were in their environment to forcibly raise the water level, thus giving them access to a drink! Schools of fish and birds who migrate great distances are actually using the positioning of stars in part to show them the way and while they are not using conventional math as we know it, they are indeed mathematical principles that are at work. Get at the cartoonist Dave Blazek for more information on Calculus trained dogs.

CH 10: NKS: The Anti-Establishment might be the Establishment!

In a time of cloning, genome mapping, and rovers on mars, it is not terribly surprising that there would be developments in science that will likely challenge our concepts and cultural artifacts as it relates to science. One of the most developed and personally awe-inspiring works comes from one man, the uber-scientist Stephen Wolfram. He has proposed a far more digital approach to science and one that would completely reposition all our sciences and how we discover new truths if it is indeed correct. While the title is perhaps a tad bland, the vision is anything but, A New Kind of Science is the idea that the behavior of very simple cellular automata can actually tell us all we need to know about the world around us. This science would not be possible until now though because it requires the horsepower of the digital world to process and simulate the results that occur. The main thrust of his premise is that from very few and very simple rules for a system, incredibly complex, unexpected and marvelous results can occur. For example, all the flight patterns of birds in flocks are governed by 3 simple rules. This lends much credence to the idea that it takes very little initial conditions to create enormous complexity. The implication of this approach is that it requires approaching all the science we do through a different lens. We can use the digital world to run simulations for different scientific scenarios and actually in the process learn about different models of the universe and ways to even create life. Regardless of how much of a restructuring occurs in science over NKS, I think it is safe to say revolutionary ideas and applications will result. Get at Wolfram Alpha (Chapter 54) for more information.

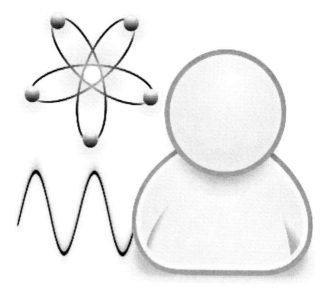

CH 11: 42% of statistics are made up!

In our modern media and culture, the bandying about of statistics to compel in arguments is a very common practice. Whether it is a politician trying to convince his electorate, or a talking head trying to make a point, we are constantly inundated with statistics aimed at quantifying the world around us. The truth is that too often statistics are misleading and misinterpreted. Let's consider a few examples. There is "shooting the barn" statistics, where you collect data without first determining the results you seek. It gets its name from the metaphor of someone shooting a bunch of arrows into the side of a barn and then circling the area with the most arrows and deeming that the target, a classic case of putting the cart before the horse. Another example is when a company achieves certain revenue and then sets their projections around these actual numbers. Another example of a flawed approach is Sample Trashing, when perfectly good data is thrown out because it does not conform to what is trying to be proved. It is common for purported Psychics to use this approach to throw out all their mistakes while highlighting anything they happen to get correct. There is also the Statistical Brick Wall, where the number in use cannot be verified because the statistical data does not even exist! A great example of this fallacy in play is when scientists predict the annual number of species that go extinct each year. The number is always arbitrarily high because the scientists are taking into account all the species that humans have not discovered yet, which clearly is a number that cannot be verified! We also need to avoid the condition of "average thinking" where someone thinks if you flip a coin ten times and it comes up heads nine of those times that somehow the next flip should be higher than 50% to obtain tails! There are also examples of very misleading using of numbers, like when a company says their product is "99.44% pure", in some cases this is a trademarked phrase and not a mathematical fact! In very close elections, it is actually possible to create scenarios in which either candidate is the winner. This is why it is so important to decide how votes are counted and how a winner is decided before the election.

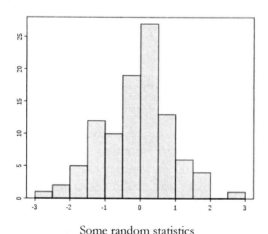

Some random statistics

CH 12: Undercover Mathematicians

While there appears to be some degree of belief that many of the people under the bright lights are vapid and devoid of geek credibility, there are in fact many examples of people in the entertainment field that excel also in the Abelian field. OK, that was a shameless math joke but let's take a second to recognize these personalities and what makes them mathematically nerdcore.

1. Winnie Cooper from the classic television show the Wonder Years, real name Danica McKellor, picked up an undergraduate degree in mathematics from UCLA and has found time to write a few well received math books aimed at teenage girls. She is also a great example of the elusive beauty and brains! Get at Stuff magazine for more information.

2. Brian May, the lead guitarist for the rock band Queen, first graduated with honors from Imperial College London with degrees in mathematics and physics. More than 30 years later, he finished his research and obtained a PhD in Physics! That is some persistent chap!

3. Art Garfunkel, half of the legendary duo, Simon &, has a Masters in Mathematics from Columbia University.

4. Who says tough guys can't be super smart? Daniel Grimaldi, an actor from the hit show the Sopranos, holds a bachelors degree in mathematics from Fordham, a masters in operations research from NYU, and a PhD in data processing from City University of New York. In fact, as of this writing, you can take classes from him in the Department of Mathematics and Computer Science at Kingsborough Community College in Brooklyn, New York.

5. David Robinson, one of the greatest players ever to tie up shoelaces for the NBA, also scored a 1320 on the SAT and then went to the United States Naval Academy to get a degree in mathematics.

6. David Dinkins is the first African American Mayor of New York, and perhaps not surprisingly has a degree, with honors, in mathematics from Howard University.

7. Tom Lehrer, the revered songwriter and parodist, is also brilliant, earning a degree in mathematics from Harvard at 18, and then followed it up with a Masters a year later!

8. Frank Ryan led the Cleveland Browns to a NFL championship in 1964 but perhaps is even better remembered for being the only player in NFL

history to hold a PhD in mathematics, from Rice University.

9. Paul Wolfowitz has been a Deputy Secretary of Defense, President of the World Bank but started with a bachelor of mathematics from Cornell.

10. Angela Merkel is the first female Chancellor of Germany, speaks fluent Russian, but first studied Physics at the University of Leipzig.

11. James Harris Simons is one of the most successful hedge fund managers ever, and subsequently one of America's richest citizens. The Financial Times has called him the "smartest billionaire" and before he ever made money, he was getting degrees from MIT and Berkeley in mathematics.

12. Bram Stoker, the author of the classic horror novel Dracula, first scared people by earning a degree in mathematics with honors from Trinity College in Dublin, Ireland.

13. William Perry has served as the United States Secretary of Defense under Bill Clinton. Before that he was a PhD in mathematics from Pennsylvania State University.

14. Paul Verhoeven has helmed classic Hollywood movies like Total Recall, Basic Instinct and

Robocop. He also is a math and physics whiz with degrees from the University of Leiden in the Netherlands.

15. Larry Gonick is a well respected cartoonist, who received an MA in mathematics from Harvard

16. Masi Oka, Hiro on the television show Heroes, doubled majored in math and computer science at Brown.

Mathematical Honorable Mentions:

Wil Wheaton of Star Trek: The Next Generation, is a pioneer on the Internet with major computer geek cred. Obviously Bill Nye the science guy, he is a mechanical engineer and can even boast to having had Carl Sagan as a professor. Phil Bredesen, the current Governor of Tennessee, has a degree in Physics from Harvard. Lisa Kudrow has an Emmy award for her acting and a biology degree from Vassar. Rowan Atkinson, aka Mr. Bean, has a masters in engineering from Oxford. Hustle and Flow's Terrance Howard is a chemical engineering degree holder from Pratt University. Frank Capra, the acclaimed director, was a Caltech graduate. Herbie Hancock is a bonafide electrical engineer. Tom Scholz, lead singer of Boston is an MIT grad. Montel Williams the talk show host is also an engineer. Dexter Holland, lead singer for the Offspring, has a bachelors and masters in molecular biology from USC. Dylan Bruno, who plays Colby on Numb3rs, has a degree in environmental engineering from MIT. Greg Graffin formed Bad Religion and also managed a PhD in Biology from Cornell. Weird Al Yankovic has a degree in architecture from California Polytechnic State University at San Luis Obispo. Mayim Bialik, the star of the television show Blossom, holds a PhD in Neuroscience.

Undercover mathematician's car

CH 13: $e^{i\pi} + 1 = 0$ is heavy duty

In the circles of mathematicians, there is some agreement that this humble formula, $e^{i\pi} + 1 = 0$, referred to as Euler's identity, is of the most beautiful and greatest that has ever been produced. Stanford mathematics professor Keith Devlin says, "Like a Shakespearean sonnet that captures the very essence of love, or a painting that brings out the beauty of the human form that is far more than just skin deep, Euler's equation reaches down into the very depths of existence." So what is it about this formula that is so captivating and powerful? For starters, it is likely to have come from the mind of Leonhard Euler, who is among the most prolific and gifted mathematicians to ever have lived. The main reason for its beauty is in its incredible ease in using so many different operations and powerful mathematical constants simultaneously. It uses the operations of addition, multiplication and exponentiation exactly once each. It uses the ubiquitous 0, 1, π, e and i mathematical constants also exactly once each. To use such ubiquitous and powerful constants in such a compact form is truly an amazing achievement!

$$e^{i\pi} + 1 = 0$$

CH 14: Economic implications of Gaussian Copula Functions

It is hard not to be aware, whether you are 18 or 80, that in recent times this great country of ours has suffered some degree of an economic meltdown. Furthermore, like so many things in modern society, the issues surrounding this financial collapse are complicated ones, making it difficult to hone in on its root cause. According to a lot of people though, we needn't look further than the Gaussian Copula Function for answers. Apparently this function convinced some banks, bond traders, insurance companies, hedge funds and other Wall Street big wigs to assess risk in an altogether risky way. It gave people with tremendous power in the financial world the ability to create correlations between seemingly unrelated events when in fact there was very little of a relationship to be found. This led to credit rating agencies becoming convinced that toxic mortgages were in fact AAA rated. This fuzzy math ignored realities and innate instabilities that are present in the mass markets. Combined with greed, and ignorantly applying mathematical principles, the financial markets as we knew them were destroyed. Perhaps an important lesson to take away from this economic collapse is that when we foster a society that elevates, appreciates and demands math fluency, we make it harder for this kind of rampant abuse to occur.

CH 15: A Proven Savant

Many can recall the classic scene from Rain Man where Dustin Hoffman's character is able to compute the number of toothpicks that had just fallen to the ground. This notion of incredible calculation or otherwise genius ability is a very profound concept. Even though of those with autism, a small minority will likely have what is termed genius abilities, it is still worth mentioning that some humans are capable of seemingly extraordinary feats of mathematical capabilities. What I think is worth investigating is how they do it and how ordinary folk might similarly tap into these skills in our brains. Consider the high functioning savant, Daniel Tammet. He recited 22,514 digits of Pi, whose digits follow a totally random sequence, in front of cameras. The prodigious memory of Kim Peek, who can effortlessly recall any content from over 12,000 books he has read. Consider Stephen Wiltshire who is able to draw a near perfect landscape, needing only a minute to memorize all the intricate details. Tony DeBlois, a blind musician, can play over 8000 songs from memory. The point of all these examples of brilliance; the ability to multiply huge numbers as quickly as a calculator, to remember and recall anything, to hear and then play music perfectly, to draw incredible detailed renderings that border photo-realistic, to read books in the time it takes to turn the page, these are all "human" abilities. Clearly, there is a different wiring in the brain that is causing such abilities, but what is incredible is that these all fall within the purview of human abilities, even if they are exceedingly rare. Future research and investigation might unearth ways and means for the ordinary person to tap into these abilities. It also begs the question, what other incredible skills do we all have the potential to do? It is from these savants and through scientists the future might reveal. Speaking of being able to tap into these abilities, the story of Rudiger Gamm is a rather interesting one. A prodigious human calculator, he is able do complex calculations instantly, but remarkably only gained his abilities in his early twenties. He also does not exhibit Savant traits, indeed it is postulated that he developed his skills through his genetics. If this is true, it definitely means that more people will be developing more "genius abilities" in the future.

CH 16: E8 and the history of the TOE

It appears that the most famous genius to have lived, Albert Einstein, actually spent the majority of his life frustrated at how to reconcile his powerful theories of Relativity with those of Quantum Mechanics. While his theory was remarkably accurate in describing physics of the very large, its equations could not work in conjunction with the very small. From this inconsistency has spawned the quest of generations of brilliant scientists to determine what physical theory can explain physics of the both large and small. This is referred to as a Grand Unified Theory (GUT) or a Theory of Everything (TOE). For several decades now, despite several attempts, no one has been able to convince the majority of scientists that they have worked out a correct TOE. In the fall of 2007, Garrett Lisi, a PhD in Physics and at the time, a relative outsider and unknown, proposed his "An Exceptionally Simple Theory of Everything". This is an attempt to solve the elusive TOE and he does so using a decidedly mathematical construct, namely the E8. E8 is perhaps the most complicated structure known to man and according to Dr. Lisi, might hold the answer to everything. While Dr. Lisi maintains a realistic and grounded opinion of his work, the String Theory establishment is not all enthused about his proposal, as it would call into question a great deal of the validity of their work.

The Large Hadron Collider (LHC), the world's largest particle accelerator, could shed light onto the theory through the discovery of new particles which Dr. Lisi's theory predicts exist.

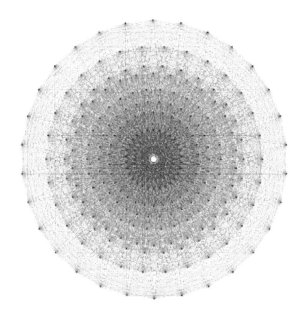

A Dramatic oversimplification of E8

CH 17: One heck of a ratio

Π or Pi is known by most people as being 3.14. Another group knows it to be the ratio of the circumference of a circle to its diameter. Another way to consider Pi is that every circle, regardless of its size, has a little over 3 times the distance around the circle compared to a line that bisects the circle into two halves. While this might seem like a rather esoteric mathematical tidbit, it has profound implications and a storied history. For starters, Pi is likely the most well known ratio ever known by humanity. Pi is defined as an irrational number. This means it is impossible to ever find any two integers that are a ratio of Pi. It also means that the digits of Pi follow a seemingly random sequence. No one to date has been able to determine if the digits of Pi end or if there is any order to them. Today, with the power of computers we know more than a trillion digits of Pi and in fact, memorizing digits of Pi is a bit of a geek phenomenon, the record is currently held by Akira Haraguchi who memorized 100,000 digits of Pi. To put that in perspective imagine trying to memorize a book hundreds of pages long of random numbers.

CH 18: A real Mathematical Hero

Paul Erdos was a Hungarian mathematician who lived during the twentieth century. He is unique due to the fact that he never maintained a permanent residence, never married, never had kids, shunned worldly possessions and basically lived an entirely nomadic existence. He lived to be 83 years and was a mathematician from an early age. While his decision to live such a life probably has cemented his reputation as an eccentric, it also afforded him the opportunity to spend his long life dedicated entirely to mathematics and in the process become the most prolific mathematician ever in history as of this writing. In fact, he has collaborated with so many different mathematicians that the Erdos number was born. This is a number that refers to how many degrees of separation any given mathematician is from working on a math paper with Erdos. So if you are Erdos you maintain the only zero and if you worked with Erdos, you have an Erdos number 1 and if you worked with someone who worked with Erdos you have an Erdos number of 2. This procedure continues, in just the same way that the classic "Six degrees of Kevin Bacon" game works. Get at Oracleofbacon.org for more information.

CH 19: Casinos love Math

The modern Casino is a truly intense form of applied mathematics at work. All Casinos trust entirely on the certainty of mathematics to generate and ensure they are viable businesses with profit margins. From the increasingly sophisticated slot machines to the action one finds at the Craps table, all games are mapped and analyzed in order for the Casino to establish confidence in being able to make money, aka the house odds. What makes Casino games differ is the way in which the element of chance figures in each game. There is a great deal of mathematics to analyze behind each game and each reveals interesting consequences. For example, do you know which game gives you the best chances to win? It is actually Craps, but not surprisingly there is a great number of rules that one must abide by in order to ensure you have access to those odds. A game that requires far less in terms of learning rules is the game of Blackjack. When you play with little errors your odds are just off being a coin toss. An example of an error is not hitting on your 16 when the dealer is showing a face card. One of the greatest nexuses of mathematics and chance is the card game Poker. Poker is a mathematically intense game, and has many examples of surprising results. Consider the chances of getting a royal straight flush in the Texas Hold'em version of Poker. There exist only 4 chances out of 2 million plus total permutations.

Poker is a game where you want to become as acquainted as possible with the probability of various events occurring. You want to first consider your starting cards and how good they are; part of that decision is also how many players there are, and where you are in the blinds. You then must reevaluate that probability after each event, like the flop, turn, and river, as well as the number of players. Certain hands, combined with the cards on the table, ensure that you are the winner, in this infrequent and powerful position, the mathematics is complete and it is up your poker persona to extract as much money as possible from the table.

CH 20: The man who was sure about uncertainty

Kurt Godel was a very influential logician, mathematician and philosopher in the twentieth century. During Godel's lifetime, there was a major attempt by the mathematical community to completely determine all the rules that govern mathematics. There was a half century of concerted efforts to figure out all the potential rules (in math we call them axioms) that form the foundations of mathematics. His genius and breakthrough was in realizing that the system of mathematics, while consistent, cannot be complete. Furthermore, consistencies in the theorem cannot be proven. This was called the Incompleteness Theorem and had profound implications for the philosophy of mathematics. It means that we can never know for sure if anything is truly correct or not, even in math! This also means that a Theory of Everything (TOE) may be even more elusive.

CH 21: More Incompleteness

One of the leading researchers who have furthered the idea of incompleteness discussed in the previous chapter is the Mathematician Gregory Chaitin. He came up with a concept he calls the Omega number. It is the centerpiece of the idea that in truly pure mathematics, there is always inherent in it an element of randomness. This fact means that all theories and concepts in math, no matter how effective or elegant, will always be tinged with an incompleteness to them. This then would rule out any permanent Theory of Everything (TOE), since a TOE is meant to be complete and the Omega number rules that out. The concept of the Omega number is related to the Turing machine. The Turing machine was a model of the first digital computer. As soon as you think about a computer program, you then must think of algorithms. An important first question when thinking of algorithms is whether any particular program is designed to stop. As was proven by Turing, there exists no test that can determine if any given program will halt or not. Now enters the Omega number which is the probability that any given program will halt. This fact makes this number irreducible, or algorithmically random. When a number is algorithmically random, that makes it maximally unknowable, and therefore infinitely complex. However, in a TOE, it must have finite complexity in order to be a theory, and the Omega number is therefore unable to be deduced.

CH 22: Have I got a Question for you!

An attempt to isolate the most difficult problems in mathematics and offer big cash prizes is certainly worth mentioning here. Both fame (not of the Hollywood type mind you!) and fortune await those clever individuals who can crack these most challenging and relevant of mathematical puzzles. There are seven problems called the Millennium Prize Problems and here they are in oh so very lay terms.

1. P vs NP (exponential time). This is perhaps the most important question in theoretical computer science. The question is whether a computer that can verify a solution in a certain time frame can also find a solution in a certain time frame. Are there questions that would take infinite time to solve with infinite computational resources?

2. Hodge Conjecture. It is a major unsolved problem in Algebraic Geometry, suffice it to say.

3. Navier-Stokes Equations. Fluid mechanics, which is applied math that deals with the motion of liquids, is immensely useful and effective. The challenge with this problem is to fill in the gaps on these insights, which still remain elusive. This would go a long way to better understanding turbulence.

4. Poincare Conjecture. The only solved problem of the Millennium Prize Problems, by Grigori Perelman, a Russian mathematician. It deals with Topology, which is the math that is concerned with spatial properties that are preserved under certain deformations, like bending and stretching.

5. Riemann Hypothesis. Perhaps the most difficult problem, it is a deep problem related to number theory, the math that is concerned with the properties of numbers. It would have implications for the distribution of prime numbers, which has profound implications in cryptography.

6. Yang-Mills Theory. This is dealing with physics, and proving that quantum field theory (you probably have heard the term quantum mechanics) is provable in the context of modern mathematical physics.

7. Birch Conjecture. Oy Vey, this is so abstract, I will leave it at that.

CH 23: When Nothing is Something

It is rather remarkable to ponder that the mathematical idea and application of zero is relatively new. Only in the 6th century AD do we see the first proof of civilization using the zero. Prior to that, people struggled working with numbers, particularly very large numbers, as the difference between a number like 15 and a number like 105 would be much harder to establish. Even the mighty Greeks, who held numbers and mathematics in high esteem, struggled with zero as a number. They wrestled with the philosophical idea that nothing could be something, and these questions became deep religious questions, even many centuries later. In 9th century India we see the first practical use of the number zero, in that it was treated as any other number. Even the ancient Chinese, a civilization rich with sophistication, took until the 13th century to develop an actual symbol for zero! It is easy in our modern society to take for granted the simplicity and necessity of the zero but for much of human civilization; it has been a complex quandary without an obvious solution. The absence of a number is in fact one of the most profound numbers there is!

CH 24: Think Binary

Before we delve into the concept of the binary system, the idea of counting needs to be revisited briefly. When we count our basic building blocks for any number are the digits 0,1,2,3,4,5,6,7,8,9. Every number is comprised of those numbers. Some people believe that we use 10 digits because we have 10 digits, namely our fingers and toes. Either way, there is no reason that we have to use 10 digits to count any number. In fact there is a well known numeral system at work that is probably in front of your nose every day. This is called the binary system and is used by the digital world. What this means is that everything that you see on a computer is actually at its core just values in the binary numeral system. This means YouTube videos, Facebook images of your friends and everything in between, rather amazing! Binary uses a 0 and a 1 to compute anything. To translate a regular number like 17 into binary takes a couple quick steps. First, remember that all numbers in binary are just zeros and ones. Next, think of a binary number as a series of slots, in which each slot is a power of two. The first slot is 2^0, or 1 and then 2^1 or 2, then 2^2 or 4 and so on. For each slot that is a 1, you add that slots value to all the other slots that have a 1. So first imagine some powers of 2: 64, 32, 16, 8, 4, 2, and 1 (2^0). We can arrive at any number by adding the proper sequence of these numbers. (Decimal numbers are arrived at in a slight modified method) So in this way we arrive at 10001 as being 17. When you see 17 on a computer, the computer sees it as 10001!

CH 25: Your order will take forever

A very interesting area of math deals with something called permutations. This differs from combinations in that order doesn't matter with a combination but order is everything in a permutation. Let's start with a simple example, the letters A,B,C. The permutations are ABC, ACB, BAC, BCA, CAB, CBA. This represents 6 ways to order 3 elements. Now it doesn't matter what the elements of the set are, they could be numbers, symbols, or people for that matter. So in our example the way we can mathematically arrive at the number 6 without having to write out every permutation by hand, we can use the factorial. Whenever you attach a "!" to a number it means that to arrive at the value, you have to multiply that number by each subsequent lower integer value until you get to 1. So for 3!, it really is $3 \times 2 \times 1 = 6$. This means then for a set of 4 unique letters, the number of permutations is 4! or $4 \times 3 \times 2 \times 1$ or 24. What you might be noticing is that as you go up in the number of elements, the total number of permutations grows very fast. Let's consider a situation in which you have 10 family members, arranging themselves in a line to take a group photograph. Like most families, an argument ensues and it is agreed that a photograph of every order of the family members should be taken to be fair. Assuming you have a fast camera and that everyone can move and take the next permutation of the family photograph every second, how long will it take to capture every way to take this picture? Well, from the previous explanation you probably have surmised it is 10! seconds. How much time is this? 3,628,800 seconds or 60,480 minutes or 1008 hours or 42 days exactly! This is also assuming that you never error in duplicating a previous permutation and have endless film. This is a shocking but completely sound result from the world of permutations.

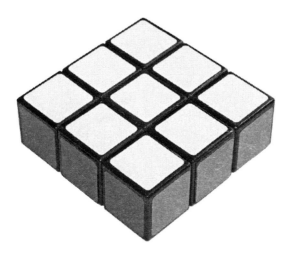

Floppy Cube, a 3 x 3 x 1 Rubik's cube

CH 26: When you need Randomness in your life

When someone says, "That was random" we generally think of it as an event without any connection to anything. The idea of unpredictability, the lack of a pattern, a process that is not deterministic, these are all traits of the term random. When we think of random numbers we tend to think of numbers that are impossible to predict from whence they came. There are many situations that model this behavior, like those lottery machines that create a fan, then the lottery balls are randomly pulled out. This works quite well for not being able to determine what numbers will be chosen. In the digital domain, that randomness is a bit harder to emulate. In fact, it is so difficult that there is a term called a pseudo-random number which refers to a number that appears to be random but is in fact not. In cryptography, which is all about how to protect information, it is very dangerous to use pseudo-random numbers to protect your data. Since there is a very deterministic process to arriving at a pseudo-random number, generally an algorithm, that process can be uncovered and therefore the information stolen. There are, however, random generators, like the Open Source Lavarnd that work by measuring noise and random.org offers *free* random numbers!

CH 27: e = mc² Redux

No big surprise, physics is tough, real tough. However a physics equation might be the best known equation known to the world, with the exception of the Pythagorean Theorem. This is Einstein's $e = mc^2$. Its direct translation is that energy (joules) is equal to the mass (kilograms) multiplied by speed of light squared. Since the speed of light is 186,000 miles per second then we know that even a very small amount of mass will contain a very large amount of energy. One of the first profound results of this formula is that mass (whether it is an ant or a skyscraper) and energy are different forms of the same things. This means that energy can be converted into mass and mass can be changed into energy. When you plug in some values, it is shocking to learn that the amount of energy in something like 30 measly grams of hydrogen is equivalent to thousands of gallons of gasoline! When extra mass suddenly converts as energy, it is called nuclear fission. This is more commonly known as the atomic bomb, which was tangible evidence of the truth and power of $e = mc^2$. Part of the reason why this is one of the most famous equations of all time is in its simplicity. In mathematics, we are always trying to consolidate as much truth into as compact a form as possible. Mathematicians like to use the adjective elegant to describe this quality of being very simple and simultaneously very concise. For

Einstein to be able to see that energy and mass are two sides of the same coin and to then use math to express this fact, and then do it in such a simple form is an intellectual marvel and the reason why you have heard of it before!

CH 28: From the Quipu to Mathematica

Perhaps an effective method to assess the level of scientific sophistication in a society is to examine the tools they use for mathematics. Many ancient civilizations consider the abacus landmark in allowing people to calculate quickly. The Quipu allowed a civilization that didn't even have written word make computations easier through the use of knots to indicate numbers. The slide rule helped scientific progress in the twentieth century. In more recent times we can certainly consider software as mathematical tools of the trade. The software Mathematica is able to compute, calculate and solve problems of depth and breadth that would have appeared like alien technology to our ancestors even a few generations ago. As further evidence of the sophistication of modern tools, scientists at IBM Research Division's Zurich laboratory built the classic math tool, the abacus, except that they did so with the individuals beads each having a diameter of about one nanometer, which is about one millionth of a millimeter! Get at Archimedes for more information.

CH 29: Through the eyes of Escher

It is quite likely that you have seen the work of M.C. Escher at some point. His artwork is fairly recognizable and typically involves impossible scenarios and tessellations (the tiling of a plane with no overlaps or gaps). He was very skilled at exploring paradoxes of space and geometry. He even wrote a paper on his mathematical approach to his artwork. He was able to bring more dimensions into the 2D of his canvas and explore ideas of infinity into his art, which result in very visually surprising effects, such as a river that seems to flow upward. He was an expert at playing with our ideas of perspective, his first print, titled "Still Life and Street" depicts a table with books and items that blend in with a street scene. Even though he did not have formal mathematical training as such, he had an incredible intuition about the visual nature of mathematics and the paradoxes that can occur. I find it rather surprising that I have not personally seen more examples of paradoxical artwork.

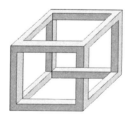

CH 30: Origami is realized geometry

Origami is the Japanese art of paper folding. It generally requires one piece of paper, and no gluing or cutting. It also uses a small number of folds that combined in different ways can create a wide variety of intricate designs. It is a very spatial and geometric art form; perhaps not surprising then it is so popular in a culture that has a very spatial and geometric language.

Because of origami and geometry's close relationship, a field has evolved, origami sekkei, which is using mathematics to determine and construct new shapes and designs, rather than the trial and error approach of earlier days. It has become a rather rigorous field, with many prominent physicists and scientists proving ever more complex designs with the benefit of mathematics and computer modeling.

CH 31: Quantifying the Physical

One of the biggest obsessions of most people is sports. The actual games vary depending on cultural roots and personal preference but regardless of choice, there is an innate desire by humans to understand and interpret games of physicality. Generally speaking, sports are considered to be a physical activity that are competitive in nature and are based on a set of clearly defined rules. Given these rules, as a game is played, the generation of statistics occurs. These statistics are the numerical results that occur from the playing of the game, for example when a basketball player makes 1 of 3 free throw shots; we now have data in the form of a percentage for this player. These statistics are critical to determining which athletes are performing well and which are not. The analysis of sports requires these statistics as objective measures to ascertain performance. With the modern reliance on all things mathematics, it is not surprising that some of those original objective measures would be challenged as truly objective and effective. This reevaluation of what metrics are best used to determine a players future performance are starting to challenge traditional measures. The stakes are high because if indeed these new metrics are able to better predict a player and/or teams performance, then that can make the difference between a winning and losing team. The most dramatic example of this shift to new measures is found in baseball, which is perhaps not coincidentally also the most statistic heavy sport. It is called Sabermetrics and helped lead to one of the biggest curses in Professional Sports to be broken, the Boston Red Sox 86 year losing streak. In particular, consider Boston Red Sox's focus on the metric of on-base percentage. New metrics are also beginning to change the interpretation in the NBA, APBRMetrics, and they are also starting to seep into the NFL and the NHL as well. The reality is that the more we can use sophisticated math in conjunction with analyzing games, the more we can distill what the true measures of success are. I expect the Metric business to only continue to reshape the way sports are interpreted.

CH 32: Geometric Progression sure adds up

The idea of exponential growth and how quickly it can grow is well illustrated in a story called the "Legend of the Ambalappuzha Paal Payasam". So the story goes that there is a king whose is so enamored by the game of chess that he offers any prize to a sage in his court if he can beat the king in a game. The sage says as a man of humble means he asks for but a few grains of rice, the specific amount to be determined by the chess board. The first square on the board will have 1 grain of rice, the second square 2 grains, the third square 4 grains, and each subsequent square having twice the amount of the preceding square. The king is rather disappointed in such a meager prize and challenges the sage for more of a substantial prize from his vast kingdom. The sage declines, and then proceeds to win the game of chess. When the king starts to count out the grains of rice, it starts to dawn on him the true nature of the sage's request. By the 40th square (there are 64 squares on a chess board), the king is in arrears on the order of a million million grains, not an insubstantial sum. By the last square he needed to procure trillions and trillions of tons of rice, more than even the world could produce. In the story, the sage morphs into the God Krishna, and the king agrees to pay the debt over time, giving out rice to pilgrims daily to pay off his debt.

CH 33: Nature = a + bi and other Infinite Details

One of the most powerful areas of mathematics is surely geometry. This puts the visual and spatial into numbers, creates order behind shapes, and has allowed almost all of modern society to emerge. For many centuries, the only geometry known and believed is what is now called Euclidean Geometry. This was named after the Greek geometer Euclid. The fundamental quality of this geometry is the idealizing of shapes. This means that when we speak of a triangle there are certain expectations required, like the fact that the sum of the angles must equal 180 degrees. What new geometries have done is to consider the realistic conditions of nature and create geometry around this imperfect reality. This has led to the development of fractal geometry which is a better approximation of nature. For example, the branch structure of trees and the design of leafs all conform to fractal patterns. The idea of fractals is geometric shapes that repeat endlessly as you zoom in on any part of the fractal. So imagine some geometric shape and imagine that as you zoom in on said shape the shape emerges again and no matter how much or where you zoom you encounter the same shape again and again. This is a powerful result and not only does it approximate reality well, it often yields results that are very aesthetically pleasing. The key to creating fractals is the use of complex numbers, which are then plotted on the complex plane. A complex number is like regular numbers except that in addition to being a Real number, it also has an imaginary component. Imaginary numbers are those that contain the component of i, where i is equal to the square root of -1. Now you might say, wait, I thought you can't, by definition, have the square root of a number be negative. You are generally correct, but the imaginary numbers allowed for the creation of a plane that normally cannot exist.

CH 34: Mundane implications of Time dilation

If we are to believe the empirical evidence behind Einstein's theories, then it is an accepted fact that time traveling into the future is simply a matter of building a very fast spaceship. For example, if we could build a ship that traveled 99.99% of the speed of light and then you spent a day on said ship when you returned to earth the rest of the world will have move much further along the time line than you have! Despite the seemingly inexplicable nature of such a result it is not science fiction, just science fact. At an incredibly small, indiscernible level, we are all at different points in time. While it might not be obvious does not make it any less true! The most likely current example would be the human who has spent as much of their life as possible in fast machines, like an Astronaut might, versus someone who is born raised and never uses any human built machines that can move very fast, like an Amish person. If the astronaut at the end of a long life was to visit the Amish person at the end of his very long life, there would be the biggest difference in time, still completely indiscernible in terms of experiencing reality but there nonetheless. Clearly, time travel into the future is just building a much faster flying machine than what our current technology is capable of. Time traveling into the past is theoretically about traveling faster than light, in which case you are catching up with time, but is a much harder problem though, since the past has already occurred so that involves retracing steps in sand that already has disappeared. If we suspend the incredible difficulties that exist with time traveling into the past, and make the assumption that future society works out the kinks and builds such a machine, it begs a very interesting question: Are there time travelers here right now? Certainly, if a group of humans can build a functioning time machine, presumably they can also create and use technology to render themselves completely invisible to us that are here now. The fact is that we cannot rule out this scenario, even if it feels altogether impossible. Get at "Planet of the Apes" for more information.

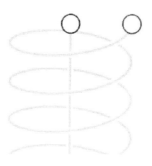

CH 35: Off on a Tangent

I figured I would veer off into an area of concern to everyone, the preservation of this little planet and all of its many inhabitants. First off, consider our current era of history where technology is accelerating at an untold and unprecedented pace, a human population that will be over 7 billion in a few years, a complex web of cultures, religions, belief systems and ever the rift between the haves and the have-nots. Such a thoroughly complicated and high stakes world is that which we live in now. It seems to me that surely the best way to ensure that humanity can handle the highs and lows of disasters and wars is to create the most international political system possible. This is the only way to transcend the country and cultural biases that seemingly have created most previous conflicts. It also would allow all countries to have a voice and create a Meta-diplomacy. Obviously, the closest manifestation of this reality is the United Nations. The problem is that the United Nations does not have enough political clout at the moment to be truly effective. So whether it is a redesign of the UN or an entirely new organization, I am quite sure that the future will require less country awareness and more world awareness to occur in order to best work together in the 21st century. Get at Carl Sagan for more information.

Ch 36: Modern Syntax Paradigms

A central tenant of math is the concept of syntax, the rules that govern mathematical systems, logic and computer programming systems. You can often find rules that apply to multiple systems, in which case the rule is generalized or the rule might only apply to one system, in which case it is a specialized rule. In the context of language, syntax can be seen as an extension of biology, since all of language and its constructs can be embodied in the human mind. This is where linguists like Noam Chomsky focus their efforts, the analysis of syntax as a means of understanding broader human behavior. One of the developments that have occurred in the digital age is the birth of new languages that use technology as a conduit. If you have ever had to decipher a text message or an Instant Message, then you were trying to understand the meaning of new syntax that is evolving. The power of mathematics can help us formalize and decipher these new forms of communication and in the process, better understand how technology shapes the way we communicate with each other. Get at UrbanDictionary.com for more information.

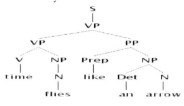

CH 37: Awesome Numb3rs

Hands down, one of the biggest contributions the entertainment industry has made to mathematics is the show Numb3rs. It concerns an FBI agent and his mathematically gifted brother. The premise of the show is that the FBI is helped in its cases by the creative utilization of sophisticated mathematical techniques. What is so terrific about this show is that the math is quite real and the cases are quite close to or based on reality. The show is well acted, well written and highly recommended! Numb3rs has probably done more than almost any other television show in history to accurately introduce correct higher level mathematics to a wide audience. The show maintains mathematical consultants from Caltech to ensure the accuracy of the math being used in the show. Some examples of the concepts that have been presented, just from Season One alone, include P vs. NP (Chapter 22), Geometric progression (Chapter 32), Monty Hall problem (Chapter 1), and Sabermetrics (Chapter 31).

MATH3MATICS RUL3S!

CH 38: A fun sampling of Math Symbols

One of the primary reasons why mathematics appears so confounding to people is because there are so many symbols that are needed to communicate any given topic. Like a foreign language, or even better, a foreign alphabet, math can appear completely alien to the uninitiated. Many of the symbols are aesthetically pleasing, and let us consider an overall categorization of symbols. Since so much of ancient mathematics stems from Greek civilization, there are many symbols that are pulled right out of the Greek alphabet. Some of the most common are zeta, beta, alpha, gamma, phi, delta, theta, lambda, and omega. There are the symbols that are most often found in discrete mathematics, these include: set, subset, intersection, union. We also find a great deal of symbols in calculus, just consider all the different ways to represent the derivative, and lest we forget the Integral symbol, mistaken for a letter s its fair share of times!

CH 39: Your consciousness can be computed

With so much progress in the sciences and in philosophy, one area that has synergy with these areas is a framework for consciousness. It is easy to look at a person sitting next to you and say they are conscious. But what about a pet dog or bees outside on a flower, or even some A.I. like Wolfram Alpha? Do these have consciousness as well, and if they do, can we then assign relative levels of consciousness? There is a theory, not proven, but worth considering, that puts consciousness in terms of information. Whether it is the streams of zeroes and ones that make up the digital world or the thoughts that emerge in your brain as you read these words, there is in both the aspect of information creation. This theory is called the Integrated Information Theory of Consciousness or ITT. It postulates that the amount of integrated information that an entity possesses corresponds to its level of consciousness. Using the language of mathematics, we can take a particular brain, consider its neurons and axons, dendrites and synapses, and then, in principle, accurately compute the extent to which this brain is integrated. From this calculation, the theory derives a single number, Φ (pronounced "fi"). The more integrated the system is, the more synergy it has, the more conscious it is. A consequence of this theory is that so many systems are sufficiently integrated and differentiated, thereby guaranteeing at least a minimal consciousness, this includes the bee, but also insects, fish and any other organism that contains a brain. This theory also does not discriminate between organic brains, like those found in a skull of a human, and the transistors, memory units and CPUs that comprise modern Personal Computers. While obviously you are not going to consider your Dell laptop as being conscious, it also does not have a null value for Φ according to the ITT!

CH 40: Auto-didactic Ivy Leaguer

The Internet is obviously a revolution of information, and is fundamentally altering the way we receive, create, understand, purchase, trade, interpret, collect, and enjoy information. There are many examples of online education, but in recent times, the quality, breadth and accessibility of this content has gone up dramatically. This is perhaps nowhere better exemplified then the OpenCourseWare project at MIT. This is a website where you can access anytime, almost the entire MIT catalog of classes. Under each class you get varying degrees of detail but almost always the actual syllabus, assignments, and required reading. In some cases you even have access to YouTube videos of all the classes taught by the instructor. In this situation, the only difference between you and a real MIT student is you just can't raise your hand for clarification. And with the video lectures, I could even pause the professor to get a snack or go do something, try doing that in real life! The site is designed well; no surprise considering this is MIT. What is perhaps most remarkable about this venture is that it is entirely free. OpenCourseWare is truly one of the greatest examples of the way in which the Internet is leveling the playing fields between the haves and have nots. For anyone that has dreamed of having an Ivy League education, OpenCourseWare is the closest thing to manifestation. While you cannot receive actual credit or a diploma, you can create for yourself or learners you are instructing a very close model of what a MIT education entails. For teachers the world over, it provides an amazing template for plugging and playing the MIT course structure for many classes. It also works great as an accompaniment to a class. Get at Academic Earth for more information.

CH 41: Zeno Paradox in time and space

If you have ever wanted to consider the paradoxical nature of infinities, look no further than Zeno's Paradox. Zeno was a Greek Philosopher who posed a set of intractable riddles that illustrate effectively the paradox of infinity. This paradox is so confounding that in a sense it uses math to imply no one can ever get anywhere! Let's dive into the specifics to see what I mean. Start by thinking about the classic problem of trying to get from point A to point B. The points themselves do not matter so it can be from wherever you are to the nearest door or Philadelphia to New York City, to offer two examples. Now, when you think about traversing this distance, it is a simple exercise to imagine that in order to cover this distance you first must go half of this distance. Now, imagine that of the remaining distance you have to go, you go half of that. You will continue to go half of each new remaining distance. This can be represented as the sum of the series $(1/2 + 1/4 + 1/8 + 1/16 + ...)$. This series is an infinite number of ever smaller values. But how can you go an infinite number of distances in finite time, regardless of how small those distances might be? While a branch of mathematics called internal set theory has come close to resolving the paradox, it remains a clever illustration of the problems inherent with infinity in a finite world.

CH 42: I needn't say more

Well, it's "The Answer to Life, the Universe, and Everything".

CH 43: I can see the past as it were

When you try to imagine the fastest thing known to man, what would you surmise that to be? Well, it is generally accepted that the speed of light has that honor. It is clocked in at 186,000 miles per second which is roughly going from the east coast to the west coast of America 60 times in one second! Clearly this is a hard-to-comprehend rate of travel but consider the relative sloth of the speed of sound in relation. The speed of sound depends on the substance in which it travels, but through the air it generally goes less than a quarter of a mile per second. So light is many hundreds of thousands of times faster than sound! When we talk about distances in the universe, which are so often incredibly long distances, it makes sense then to use the speed of light in describing the distance. For example, the distance from our sun to earth is about 8.5 light minutes, which is the distance that light will travel in 8.5 minutes. Considering what it can accomplish in one second, 8.5 minutes is a very long way. What makes this distance paltry in comparison is considering the brightest star in the sky, Sirius. This star is approximately 9 light years away. This means that the Sirius star/sun is the distance from the earth that it would take light to travel if it was on a straight line with no breaks for 9 years! This is an incredible distance in the context of our lives. It also means that when we gaze at Sirius on any given night, we are actually seeing it as it was years in the past simply because there is no way to view the light of the star as it is at the moment you are staring at because it takes light too long to show up, 9 years in the example of Sirius. With the latest telescopes we can even peer at stars that are totally invisible to the naked eye and subsequently much further a distance then a super close star like Sirius. For example, using Hubble, we have been able to look at the light of stars from near 20 billion years ago! This is very close to the big bang, which is currently purported to be the beginning of time. So if ever you long to gaze into the past, you needn't look further then the sky at night to see another place *and* time. Considering the profundity of time, I think star gazing is a rather remarkable and ignored fact of nature.

CH 44: Music is a lovely triangulation of sorts

It is rather easy to persuade most people on the power of music, its evocative and subjective powers are incredibly visceral and can reach across cultural and other man made constructs to activate feelings across the spectrum of human emotions. Not surprisingly, mathematics and music go hand in hand. In fact, all of music is a fantastic lesson in applied mathematics. From the timing of a drummer to the frequency space of octaves to the Fibonacci number appearing in musical works, music is in a sense the turning of random sounds and random timing to a mathematically pleasing order to our ears. There are even scientists who are dedicated to analyzing the synergies between math and music and making discoveries into these connections. There is a free program called Chord Geometries that plays chords in different 3D environments. Also, get at Wolfram Tones for more information.

CH 45: Mobius Strip: Assembly Required

The Mobius strip is a classic example of a physical paradox. It is created by taking a strip of paper, twisting it, and then taping or gluing the ends together. The result is a surface that in mathematics is called non-orientable. If an ant was to be placed on it and walked all the way around the strip, the ant will have traversed both sides of the strip of paper without ever having to cross an edge of the paper! You can make your own Mobius strip by taking a thin strip of paper, twisting it once and then taping the two ends together and voila, a mathematical paradox!

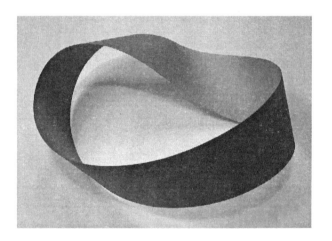

CH 46: I will never use this

One of the most common phrases that a math teacher is likely to hear is the classic, "Why are we bothering to learn this, I will never use this in real life." The truth is that while a great deal of mathematics you learn may not be explicitly used later in life for most of the population, the truth is that you learn it primarily as a means of education to the ends of exercising your brain. When you hear someone taking a math class complaining about how irrelevant it is to their lives, remind them it is fantastic for keeping your brain strong! Besides the mental exercise aspect, it is no small fact that our entire world runs on numbers, applied though it may be. Consider the cash register at the local deli to the scale in your bathroom to the taxes you do every year to buying some gas to the receipt for anything you purchase to your phone number to your favorite team's sports statistics to weather predictions to how much food to buy for dinner to poker night with your buddies to calculating the tip for the great service in your favorite local restaurant to playing video games to anytime you count, measure, compare values to channel surfing to your address, geographic or digital (IP) to your watch to the calendar on the wall to infinity and beyond! Get at your neighborhood math teacher for more information.

CH 47: Choice Words

Here are some of my favorite quotations related to mathematics; I hope they inspire you as much as they inspire me!

James, William (1842 - 1910)
The union of the mathematician with the poet, fervor with measure, passion with correctness, this surely is the ideal.
Collected Essays.

Hilbert, David (1862-1943)
Mathematics knows no races or geographic boundaries; for mathematics,the cultural world is one country.
In H. Eves Mathematical Circles Squared, Boston, 1972.

Kasner, E. and Newman, J.
Perhaps the greatest paradox of all is that there are paradoxes in mathematics.
Mathematics and the Imagination, 1940.

Aristotle (ca 330 BC)
The whole is more than the sum of its parts.
Metaphysica 10f-1045a

Bagehot, Walter
Life is a school of probability.
Quoted in J. R. Newman (ed.) *The World of Mathematics*, 1956.

Bell, Eric Temple (1883-1960)
"Obvious" is the most dangerous word in mathematics.

Sofia Kovalevskaya
"It is impossible to be a mathematician without being a poet in soul."

Blake
What is now proved was once only imagin'd.
The Marriage of Heaven and Hell, 1790-3.

Carroll, Lewis
"Alice laughed: "There's no use trying," she said; "one can't believe impossible things." "I daresay you haven't had much practice," said the Queen. "When I was younger, I always did it for half an hour a day. Why, sometimes I've believed as many as six impossible things before breakfast."
Alice in Wonderland.

Descartes, Rene (1596-1650)
omnia apud me mathematica fiunt.
With me everything turns into mathematics.

Darwin, Charles
Mathematics seems to endow one with something like a new sense.
In N. Rose (ed.) *Mathematical Maxims and Minims*, 1988.

Disraeli, Benjamin
There are three kinds of lies: lies, damned lies, and statistics.
Mark Twain. Autobiography

Dyson, Freeman
For a physicist, mathematics is not just a tool by means of which phenomena can be calculated, it is the main source of concepts and principles by means of which new theories can be created.
Mathematics in the Physical Sciences.

Einstein, Albert (1879-1955)
Gott wurfelt nicht. (God does not play dice)

Hawking, Stephen Williams (1942-)
God not only plays dice. He also sometimes throws the dice where they cannot be seen.
Nature 1975.

Pope, Alexander (1688-1744)
Order is Heaven's first law.
An Essay on Man IV.

Einstein, Albert (1879-1955)
Since the mathematicians have invaded the theory of relativity, I do not understand it myself anymore.
Albert Einstein, Philosopher-Scientist, Evanston, 1949.

Hofstadter, Douglas R. (1945 -)
Hofstadter's Law: It always takes longer than you expect, even when you take into account Hofstadter's Law.
Godel, Escher, Bach 1979.

Kepler, Johannes (1571-1630)
Where there is matter, there is geometry.
(Ubi materia, ibi geometria.)
J. Koenderink *Solid Shape, 1990.*

Locke, John
...mathematical proofs, like diamonds, are hard and clear, and will be touched with nothing but strict reasoning.
D. Burton, Elementary Number Theory, 1980.

Newton, Isaac (1642-1727)
Hypotheses non fingo.
I feign no hypotheses.
Principia Mathematica.

Poincare Jules Henri (1854-1912)
Thought is only a flash between two long nights, but this flash is everything.
In J. R. Newman (ed.) The World of Mathematics, 1956.

Pordage, Matthew
One of the endearing things about mathematicians is the extent to which they will go to avoid doing any real work.
In H. Eves Return to Mathematical Circles, 1988.

Seneca
If you would make a man happy, do not add to his possessions but subtract from the sum of his desires.
In H. Eves Return to Mathematical Circles, 1988.

Tolstoy, Count Lev Nikolgevich (1828-1920)
A man is like a fraction whose numerator is what he is and whose denominator is what he thinks of himself. The larger the denominator the smaller the fraction.
In H. Eves Return to Mathematical Circles, 1989.

CH 48: **Latest in building marvels!**

A terrific example of how the latest in mathematics can appear to us in our modern world is found in architecture. Using the latest in engineering and architectural know-how, buildings are rising all over the world with some of the most surprising and unique designs that could be fathomed. Consider the Guggenheim, the Gherkin in London, Ras Al-Khaimah Gateway in the United Arab Emirates, the Rotating Wind Power Tower in Dubai, the National Stadium in China, (also known as the Birds nest), the Dubai Towers and the Freedom Tower on the site of the World Trade Center. All these places could not have been built until recently due to their sophisticated designs, which make heavy use of applied mathematics. They are all easy to admire and gaze at and every last one uses the latest in mathematical understanding to achieve their stunning forms.

30 St Mary Axe in London AKA the Gherkin

CH 49: Your eyes do not tell the whole story

Optical Illusions are at their core a manifestation of the different properties of light and how these principles can fool the optical systems that we use to make sense of the world around us. Most of the mathematics of these illusions is really applied math, so knowing physics can help clarify how these illusions actually trick us.

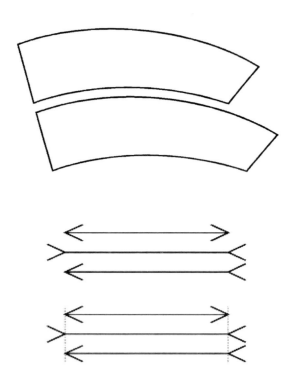

CH 50: This Internet Cred is worth listening to

DARPA, which stands for the Defense Agency Research Projects Agency, is an agency in the Department of Defense that is responsible for the development of new technologies that can be used by the Military. It was originally formed in 1958 when it became clear through the launching of the space shuttle Sputnik by the then USSR that competition could rapidly harness military technology to great effcct. DARPA is largely credited with help ushering in the development of the Internet. It is not terribly surprising that DARPA is seeking innovative research proposals in the area of mathematics with the small goal of "dramatically revolutionizing mathematics". Specifically, DARPA had identified 23 mathematical challenges that have the potential to be profound mathematical breakthroughs. Let's consider a few of the most interesting. Challenge One is to build a mathematical theory that can successfully build a functional brain, not using biological constructs but pure mathematics to recreate the brain. Challenge Nine is trying to determine whether we can build novel materials using breakthroughs in our understanding of three dimensions. Challenge Twelve is about using mathematics to understand and control the strangeness of the quantum world. Challenge Twenty Three is to determine the fundamental Laws of Biology, something that would likely involve determining the mathematics for other challenges first, like Challenge One. Based on these challenges and its track record, I surmise it a safe bet that DARPA will continue to be at the forefront of progress and mathematics in the 21st century.

CH 51: The A.I. Inflection Point

One of the most intriguing and thoroughly overlooked organizations could end up being the Singularity Institute. Their primary goal, esoteric as it might sound, seems to distill down to seeding good A.I. You are probably asking what on earth is that suppose to mean. Well, A.I. refers to Artificial Intelligence, which we know can be a rather broad term. After all, there is A.I. in vacuum cleaners, cars, planes, toys, to name but a few. In these common place examples, the A.I. has a very narrow domain of abilities, but excels at those abilities. A.I. is weak in a parallel processing context, where it is about being able to tie together very disparate concepts effectively, and where humans really excel. However, it is reasonably assumed that over time, whether it is minutes from now or decades from now, the software underlying certain Artificial Intelligence will have in its code the recursive ability to improve on its own code. This means that future A.I. will actually be able to self improve on its own! This remarkable likelihood means in certain theories that A.I. could very quickly develop the ability to more efficiently and effectively identify and solve problems than humans do. If such an event does indeed happen, it would mark a singularity event since the brain of such A.I. could be more effective than humans and since we can't predict outside of our own capabilities, we would have no idea of what such a future would look like, hence the term singularity.

The purpose of the Singularity Institute is to do all it can to lay the foundation and conditions for the best possible good A.I. to evolve and therein not have any desire to destroy mankind. While this idea seems more science fiction than fact, this organization is firmly rooted in mathematics and science. It is just that such a topic is so intellectually arcane that most people have more pressing things to consider, rather ironic considering the stakes if the good people at the Singularity Institute are indeed correct! The key to their success is mathematical breadth and depth and the ability to convert this knowledge into practical good A.I. developments. Ultimately, the language of math will best describe the A.I. that this organization seeks to build. Get at the Singularity University for more information.

CH 52: I know Kung Fu

Today, there are many different forms of self defense, from Kung Fu to Akido to Karate to Taekwondo. These art forms are popular the world over, and help their practitioners develop focus, concentration, relaxation and self confidence. Since these systems are based on very specific patterns of movement, we can analyze these movements with mathematics to better shed light on why they are effective. For example when we consider strikes, which every art form possesses, we can determine the energy of the strike. Using the mass of the punch, gravitational acceleration, the velocity of the fist, and depending on the punch, the torque velocity of the fist, we can effectively calculate the overall energy of a given strike. Not surprisingly, the amount of mass has a linear relationship with energy so more mass equals more energy. Interestingly, a shift in the overall height of the body from the beginning of the strike to its end also has a linear relationship with energy. However, more than these factors, speed has a quadratic relationship with energy so the faster the strike the greater the energy. This is good news for people with low mass, they can more than compensate by an increase in overall speed relative to their higher mass opponent. Based on this understanding, it makes sense then that kicks can deliver a greater mass and in many cases a greater velocity, which gives kicks higher overall energy to strikes involving the arms.

Besides the ability to quantify energy from strikes and kicks, we can also look at Martial Arts in a strict geometric sense, which can help analyze the efficiency of the movements involved. For example, most people know that the shortest distance between two points is a straight line. This fact is an example of how the art form Wing Chun is able to use efficiency to be effective. Wing Chun was also invented by a nun, and allows a much smaller opponent to beat a much bigger one. Disclaimer: I am a disciple of the art of Traditional Wing Chun Kung Fu. Speaking of effective, the best way to be so in any art form is one word: practice, and lots of it. The modern American is used to instant gratification but the true reward in martial arts is a lifetime of practice to *start* to learn how to do it correctly!

CH 53: We eat this stuff up

When you talk about cooking, baking and anything to do with food preparation, you don't have to go far until you run into the world of mathematics. Cooking is really a combination of art and science, the art part being that so much of enjoying food is a subjective experience. However, whenever you use a recipe, you are using precise measurements to ensure you get the result you seek. In addition, the length of time, the temperature, the number of ingredients, and a whole host of other details, all are based on simple math to create the desired result. On an unrelated note, did you know that you cannot physically break a piece of uncooked spaghetti in half! No matter how often or hard you try, you are guaranteed to always result in at least 3 pieces, the math behind this strange fact is in fact rather complex.

CH 54: Alpha Behavior

One of the most sophisticated websites on the Internet in my estimation can be found at www.wolframalpha.com. It is the brainchild of Stephen Wolfram and his team at Wolfram Research. (You can also read about him in Chapter 10). The ultimate goal of Wolfram Alpha is to accumulate and organize all of humanity's knowledge, and then through a very simple Google-like interface allows anyone, anytime access to this knowledge for free. You might be asking how such an idea is any different than what the folks over at Google are up to. The answer is revealed in the method by which these two websites find information. With Google, it is a matter of indexing pages and then having an algorithm that relevantly lists pages that are based on the keywords that the user supplies. In the case of Alpha, it is about taking the input and actually performing mathematical computations and calculations. This is a more technically challenging problem to crack, but it is also a more powerful application. For example, you can type in distance from the earth to the moon, and Alpha will actually perform differential equations to instantly calculate our distance from the moon at the moment the query was made. You can enter your birthday and immediately know your age in months, days, etc. You can type your birthplace and receive a detailed weather report on the day you were born. It can perform integrals (calculus based formulas for determining areas around functions), it can have a sense of humor if you ask it how it is feeling, it can create nutrition labels, like the ones on the packaging of most foods but for precise quantities and types of food that you specify. The list goes on and on with the capabilities of this software and the likelihood is that Wolfram Alpha will only become more capable and effective as time goes by.

CH 55: That thing on your wrist is just a Temporal Dimension Gauge.

When you take a minute and think about it, it is rather remarkable and ordinary at the same time that we live in a very well defined space commonly referred to as space time. Hark back to algebra class where we see the x axis (a point), then the x and y axis (a plane) and then the x, y and z axis (box). It is this 3 dimensional x, y and z coordinate system that we walk around in all day long. We then add a linear, meaning moving in one direction, dimension of time and we have our 4D lives, totally mapped out. As many scientists have surmised, it is impossible to rule out the possibility, however remote, of sentient beings in our cosmos that live in a greater then 4D existence. This naturally would make the capture of such a being exceedingly difficult, as it would likely be able to disappear in front of our eyes as it moves through its own dimensions. Get at the Flatlanders for more information. The strangeness of living in time and space is it is incredibly difficult to think about without using those constructs in our thoughts. However, if science has taught us anything, is that the more we know the more we realize we do not know. This constant widening of the unknowable is almost teasing our comprehension abilities. Is it possible to be a sentient being and not live in the construct of time? Can something exist without the notions of past, present and future? Is even the idea of eternity a drastic over simplification of some larger concept? What limits to understanding greater truths about the universe and humanity exist due to our preoccupation with space and time?

56. Stream of consciousness, Bing for details

1) If you think about one of the most prestigious accomplishments any country can boast of, consider taking people to outer space and back safely, now try doing any of that without intense applications of mathematics.

2) If you like to watch modern movies then you like visual effects and if you like visual effects then you like a great deal of applied math.

3) Nowadays, when you use the Internet, there are many reasons that demand the need for security and this encryption and protection of data is entirely based in mathematical formulas.

4) While there might be a thrill associated with playing the lottery, most give practically no chance for winning, for example the popular SuperEnalotto in Italy, requires the player to match 6 numbers out of a possible 90, the odds for such a feat are about 1 in over 620 million.

5) Manhole covers are round, and the reason is that any other shape would allow the cover to fall through the hole at the right angle, with a circular design there is no such worry, and the circle is an efficient use of area.

6) There even exists a mathematical approach to finding your spouse, first determine the number of total likely partners you are to have, then divide that number by e, which is roughly 2.72, then after you have had that number of partners you should pick the next partner that exceeds all the partners up until that point, this gives you a 37% of picking your best mate, which is the highest certainty you can guarantee for yourself.

7) There are a growing number of artists that consider themselves "nerdcore", generally hip hop music but not always, and oftentimes they integrate math lingo into their rhymes, here are a few artists with math metaphors that come to mind: funky49, MC Plus+, MC Hawking, Deltron 3030, Monzy and YTCracker.

8) If you were an older, wealthy individual with no real heirs, would you be intrigued by the idea of cloning yourself so that you could bequeath your fortune to yourself? Even if you thoroughly object to this idea, would everybody? Given the increasingly widespread knowledge of cloning technologies and the distribution and prevalence of personal wealth, there is perhaps already a secret clone population that is alive and well and growing every day. My personal guess is that right now the percentage of the world's population that are clones is about .000000042%. What could it be 10 years from now? 50 years?

9) the internet has been a wonderful place to utilize the power in numbers theory, Great Internet Mersenne Prime Search (GIMPS), Search for Extra-Terrestrial Intelligence (SETI) are all using the network to search for their respective answers.

CH 57: Q.E.D.

This acronym, which stands for the Latin "quod erat faciendum" and means "which was to be demonstrated", is a common way of ending a math proof. It really signaled when proving things in mathematics became less about assertion and more about deduction. This became common during the time of the early mathematicians like Euclid. So when you see Q.E.D. you know you have reached the end of the proof so the author better had proven his point, literally!

Bibliography Note:

All the images contained herein were obtained from searching using subject keywords on Wikimedia Commons, a public domain resource. If you have found any mistakes or you have any comments, please email me at jeff@lovemathematics.com.

INDEX:

42, 78
Abacus 59
Artificial Intelligence 93 – 94
Bacon, Kevin 43
Binary number system 52
Biology 19, 70
Birthday Paradox 14
Blackjack 44
Blazek, David 26
Architecture 89
Caltech 71
Cantor Transfinite Numbers 16
Casinos 44
Celebrity 31-35
Chailtin, Gregory 47
Chemistry 19
Chess 64
Chinese 51
Chomsky, Noam 70
Consciousness 73-74
Craps 44
Crow 26
Cryptography 56
DARPA 91-92
E8 40
$e = mc^2$ 57
Education 75- 76, 83
Erdos, Paul 43
Escher, M.C. 60
Euler 36
Food 97
Fractals 65-66

Gaussian Copula 37
Genius 38-39
Geometry, Euclidean 23, 65
Godel, Kurt 46
Greeks 51
Hardware 10
Humor 20-22
Imaginary Numbers 65-66
Incompleteness Theorem 46
Infinity 16, 77
Infinite Monkey Theorem 5
Internet 91
Jokes 20-22
Jourdain's Card Paradox 18
Large Hadron Collider 41
Legend of the Ambalappuzha Paal Payasam 64
Light Years 79-80
Lisi, Garrett 40
Mathematica 59
Mathers, Marshall 16
Martial Arts 95 - 96
MC Plus + 16
Millenium Prize Problems 49-50
M.I.T. 75-76
Mobius Strip 82
Monty Hall (Problem) 12
Music 81
Natural Numbers 16
New Kind of Science 27
Numb3rs 71
Omega number 47
Optical Illusions 90
Origami 61
Perelman, Grigori 50

Permutations 54
Physics 19
Pi 36, 42
Pigeonhole Principle 15
Poker 44
Probability 12 -15
Psychology 19
Q.E.D. 104
Quipu 59
Quotations 84 - 88
Randomness 56
Real numbers 16
Sabermetrics 62
Sagan, Carl 69
Sample Trashing 29
Singularity Institute 93- 94
Software 10
Sports 62-63
Statistical Brick Wall 29
Sumerians 25
Symbols 72
Syntax 70
Theory of Everything 40, 46
Time 67-68, 79-80, 100 - 101
Topology 50
United Nations 69
Wolfram Alpha 98 - 99
Wolfram, Stephen 27, 98
Wolfram Tones 81
Zeno's Paradox 77
Zero 36, 51